11
2/07

WITHDRAWN

SPARKS OF LIFE

Chemical Elements that Make Life Possible

CARBON

by

Jean F. Blashfield

RAINTREE
STECK-VAUGHN
P U B L I S H E R S
A Steck-Vaughn Company

Austin, Texas

Special thanks to our technical consultant
Jeanne Hamers, Ph.D.,
formerly with the Institute of Chemical Education,
Madison, Wisconsin

Development: Books Two, Delavan, Wisconsin
Graphics: Krueger Graphics, Janesville, Wisconsin
Interior Design: Peg Esposito
Photo Research and Indexing: Margie Benson

Raintree Steck-Vaughn Publisher's Staff:
Publishing Director: Walter Kossmann   Project Editor: Frank Tarsitano
Design Manager: Joyce Spicer   Electronic Production: Scott Melcer

Library of Congress Cataloging-in-Publication Data:
Blashfield, Jean F.
      Carbon / by Jean F. Blashfield.
          p. cm. — (Sparks of life)
      Includes bibliographical references (p.   -) and index.
      Summary: Presents the basic concepts of carbon, one of the most important chemical elements found in all living things.
          ISBN 0-8172-5041-7
          1. Carbon — Juvenile literature. [1. Carbon. 2. Organic compounds.]   I. Title.
    I. Series: Blashfield, Jean F. Sparks of life.
    QD181.C1B57 1999   98-4517
    546' .681--dc21   CIP
                                                                                                    AC

Printed and bound in the United States
1 2 3 4 5 6 7 8 9 LB 02 01 00 99 98

PHOTO CREDITS: Archive Photos 11; Marjorie Benson 35; B.I.F.C. cover; Chicago Historical Society, P&S-1964.0521–Gary Sheahan 18; Cleveland Public Library 54; Coca-Cola® 26; Corbis-Bettman 23; ©Robert Fried, Stock Boston 20; Ewing Galloway, Inc. 46; Field Museum of Natural History 32; Index Stock Photography, Inc. 25; JLM Visuals 13, 15; Barbara Krause 50; NASA 14; ©Harry J. Przekop, Stock Shop/Medichrome cover, 31, 40, 42; Scripps Institution of Oceanography, UCSD 28; Stock Boston 20, 43; Trek Bicycle Corporation 52; United Nations Photo Library cover; University of Pennsylvania Museum, Philadelphia (Neg. #S4-63180) 57; USDA–ARS Information Staff 36, 51; ©Charles D. Winters, Photo Researchers cover.

CONTENTS

Periodic Table of the Elements

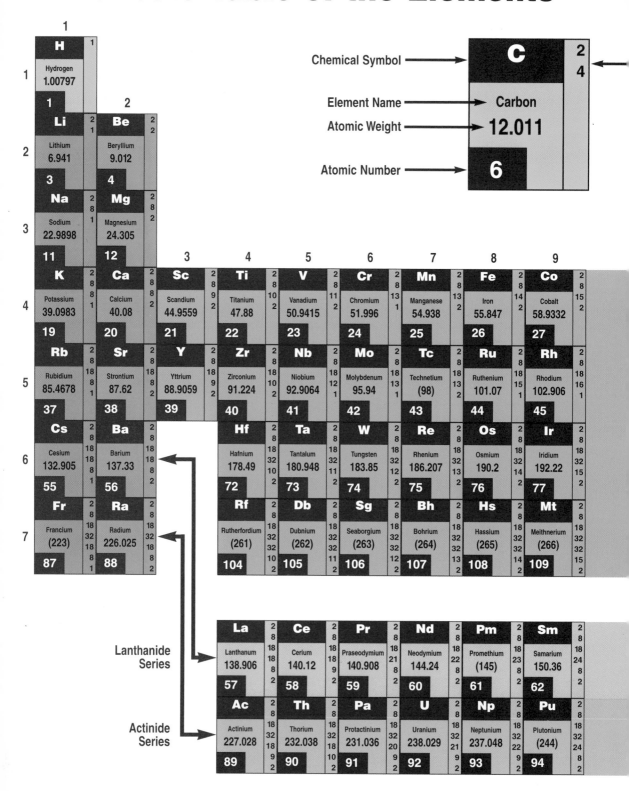

Number of electrons in each shell,
beginning with the K shell, top.

See next page for explanations.

18	
He	2
Helium	
4.0026	
2	

13	14	15	16	17	18
B 2,3	**C** 2,4	**N** 2,5	**O** 2,6	**F** 2,7	**Ne** 2,8
Boron	Carbon	Nitrogen	Oxygen	Fluorine	Neon
10.81	12.011	14.0067	15.9994	18.9984	20.179
5	6	7	8	9	10
Al 2,8,3	**Si** 2,8,4	**P** 2,8,5	**S** 2,8,6	**Cl** 2,8,7	**Ar** 2,8,8
Aluminum	Silicon	Phosphorus	Sulfur	Chlorine	Argon
26.9815	28.0855	30.9738	32.06	35.453	39.948
13	14	15	16	17	18

10	11	12	13	14	15	16	17	18
Ni 2,8,16,2	**Cu** 2,8,18,1	**Zn** 2,8,18,2	**Ga** 2,8,18,3	**Ge** 2,8,18,4	**As** 2,8,18,5	**Se** 2,8,18,6	**Br** 2,8,18,7	**Kr** 2,8,18,8
Nickel	Copper	Zinc	Gallium	Germanium	Arsenic	Selenium	Bromine	Krypton
58.69	63.546	65.39	69.72	72.59	74.9216	78.96	79.904	83.80
28	29	30	31	32	33	34	35	36
Pd 2,8,18,18	**Ag** 2,8,18,18,1	**Cd** 2,8,18,18,2	**In** 2,8,18,18,3	**Sn** 2,8,18,18,4	**Sb** 2,8,18,18,5	**Te** 2,8,18,18,6	**I** 2,8,18,18,7	**Xe** 2,8,18,18,8
Palladium	Silver	Cadmium	Indium	Tin	Antimony	Tellurium	Iodine	Xenon
106.42	107.868	112.41	114.82	118.71	121.75	127.6	126.905	131.29
46	47	48	49	50	51	52	53	54
Pt 2,8,18,32,17,1	**Au** 2,8,18,32,18,1	**Hg** 2,8,18,32,18,2	**Tl** 2,8,18,32,18,3	**Pb** 2,8,18,32,18,4	**Bi** 2,8,18,32,18,5	**Po** 2,8,18,32,18,6	**At** 2,8,18,32,18,7	**Rn** 2,8,18,32,18,8
Platinum	Gold	Mercury	Thallium	Lead	Bismuth	Polonium	Astatine	Radon
195.08	196.967	200.59	204.383	207.2	208.98	(209)	(210)	(222)
78	79	80	81	82	83	84	85	86
(Uun) 2,8,18,32,32,17,1	**(Unu)** 2,8,18,32,32,18,1	**(Uub)** 2,8,18,32,32,18,2						
(Ununnilium)	(Unununium)	(Ununbium)						
(269)	(272)	(277)						
110	111	112						

Alkali Metals	**Transition Metals**	**Nonmetals**	**Metalloids**	**Lanthanide Series**	
Alkaline Earth Metals	**Other Metals**	**Noble Gases**	**Actinide Series**	**COLOR KEYS**	

Eu 2,8,18,25,8,2	**Gd** 2,8,18,25,9,2	**Tb** 2,8,18,27,8,2	**Dy** 2,8,18,28,8,2	**Ho** 2,8,18,29,8,2	**Er** 2,8,18,30,8,2	**Tm** 2,8,18,31,8,2	**Yb** 2,8,18,32,8,2	**Lu** 2,8,18,32,9,2
Europium	Gadolinium	Terbium	Dysprosium	Holmium	Erbium	Thulium	Ytterbium	Lutetium
151.96	157.25	158.925	162.50	164.93	167.26	168.934	173.04	174.967
63	64	65	66	67	68	69	70	71
Am 2,8,18,32,25,8,2	**Cm** 2,8,18,32,25,9,2	**Bk** 2,8,18,32,26,9,2	**Cf** 2,8,18,32,28,8,2	**Es** 2,8,18,32,29,8,2	**Fm** 2,8,18,32,30,8,2	**Md** 2,8,18,32,31,8,2	**No** 2,8,18,32,32,8,2	**Lr** 2,8,18,32,32,9,2
Americium	Curium	Berkelium	Californium	Einsteinium	Fermium	Mendelevium	Nobelium	Lawrencium
(243)	(247)	(247)	(251)	(254)	(257)	(258)	(259)	(260)
95	96	97	98	99	100	101	102	103

A Guide to the Periodic Table

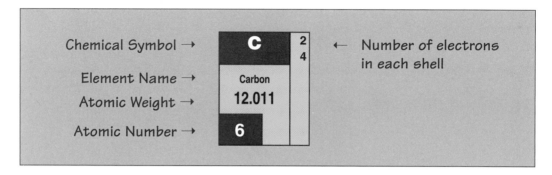

Chemical Symbol → **C**   2 4   ← Number of electrons in each shell

Element Name → Carbon

Atomic Weight → **12.011**

Atomic Number → **6**

Symbol = an abbreviation of an element name, agreed on by members of the International Union of Pure and Applied Chemistry. The idea to use symbols was started by a Swedish chemist, Jöns Jakob Berzelius, about 1814. Note that the elements with numbers 110, 111, and 112, which were "discovered" in 1996, have not yet been given official names.

Atomic number = the number of protons (particles with a positive charge) in the nucleus of an atom of an element; also equal to the number of electrons (particles with a negative charge) found in the shells, or rings, of an atom that does not have an electrical charge.

Atomic weight = the weight of an element compared to a standard element, carbon. When the Periodic Table was first developed, hydrogen was used as the standard. It was given an atomic weight of 1, but that created some difficulties, and in 1962, the standard was changed to carbon-12, which is the most common form of the element carbon, with an atomic weight of 12.

The Periodic Table on pages 4 and 5 shows the atomic weight of carbon as 12.011 because an atomic weight is an average of the weights, or masses, of all the different naturally occurring forms of an atom. Each form, called an isotope, has a different number of neutrons (uncharged particles) in the nucleus. Most elements have several isotopes, but chemists assume that any two samples of an element are made up of the same mixture of isotopes and thus have the same mass, or weight.

Electron shells = regions surrounding the nucleus of an atom in which the electrons move. Historically, electron shells have been described as orbits similar to a planet's orbit. But actually they are areas with a range of specific energy levels, in which certain electrons vibrate and move around. The shell closest to the nucleus, the K shell, can contain only 2 electrons. The K shell has the lowest energy level, and it is very hard to break its electrons away. The second shell, L, can contain only 8 electrons. Others may contain up to 32 electrons. The outer shell, in which chemical reactions occur, is called the valence shell.

Periods = horizontal rows of elements in the periodic table. A period contains all the elements with the same number of orbital shells of electrons. Note that the actinide and lanthanide (or rare earth) elements shown in rows below the main table really belong within the table, but it is not regarded as practical to print such a wide table as would be required.

Groups = vertical columns of elements in the Periodic Table; also called families. A group contains all elements that naturally have the same number of electrons in the outermost shell or orbital of the atom. Elements in a group tend to behave in similar ways.

Group 1 = alkali metals: very reactive and never found in nature in their pure form. Bright, soft metals, they have one valence electron and, like all metals, conduct both electricity and heat.

Group 2 = alkaline earth metals: also very reactive and thus do not occur in pure form in nature. Harder and denser than alkali metals, they have two valence electrons that easily combine with other chemicals.

Groups 3–12 = transition metals: the great mass of metals, with a variable number of electrons; can exist in pure form.

Groups 13–17 = metalloids, nonmetals, and other transitional metals. Metalloids (also called semimetals) possess some characteristics of metals and some of nonmetals. Unlike metals and metalloids, nonmetals do not conduct electricity.

Group 18 = noble, or rare, gases: in general, these nonmetallic gaseous elements do not react with other elements because their valence shells are full.

THE 500-POUND GORILLA

Carbon is one of the most important chemical elements found in all living things. It's also found in the stars, the sun, comets, and in the atmospheres of many planets, usually in the form of molecules of carbon dioxide. The fact that it was in Earth's atmosphere allowed life as we know it to develop. Carbon occurs in almost every chemical compound found in living things.

James R. Heath, a scientist at the University of California, has called the carbon atom "the 500-pound gorilla of chemistry." He says, "It does whatever it wants." Yet, carbon is not even one of the ten most abundant elements in the universe. Early chemists thought that only living things, or organisms, could make the special chemicals found within them. They called those chemicals organic compounds, and they called their work organic chemistry. In the nineteenth century, it was discovered that organic compounds could be made in a laboratory. They did not need any special "spark" that only life could provide. Today, organic chemistry is defined as the study of compounds of carbon.

Carbon compounds are among the oldest known. Coal and charcoal have been used for fires since prehistoric times. Centuries ago, people found that the soft, slippery form of carbon called graphite could be used for writing. The hardest of all known natural materials, diamonds, are a matchless, sparkling form of carbon.

Order from Chaos

The name carbon comes from the Latin word for "charcoal" or "ember." Ancient people learned—probably by accident—that they could burn wood in a special way to make something even more useful. Perhaps they threw wet clay over a burning fire and oxygen (O, element #8) could no longer reach it. After the wood under the clay stopped smoldering, they found that it had turned into chunks of lightweight black "stuff" that could be burned again. This black "stuff" was charcoal. It burned without smoking and made a hotter fire than wood did.

Over time, other ancient people learned that chunks of another black substance called coal could be treated in the same way as wood. The result was coke, which also burned cleaner and hotter than the original coal. None of these ancient people knew they were discovering things about the element carbon.

Belgian alchemist Jan Baptista van Helmont helped bring about the major change from mystical (and wishful) alchemy to modern chemistry in the early 1600s. He had not intended to do this, because he was truly dedicated to combining alchemy and religion. However, his experiments forged a new way of thinking about materials that eventually became the science of chemistry.

Helmont later discovered that the invisible gas coming from the fermentation of foods, from burning wood, and from seashells placed in vinegar, was all the same gas, which he called "fixed air." Only later would "fixed air" be identified as carbon dioxide or CO_2.

In the latter part of the 1700s, French scientist Antoine Lavoisier studied diamonds. He found that if a diamond were heated, a gas was given off. He eventually identified the gas as the same one Helmont had worked with. It was finally given the name carbon dioxide.

Then, in 1799, another Frenchman, Guyton de Morveau, found that heat and pressure could turn a diamond into graphite. He also knew that if molten iron had coal added to it, the iron turned into steel. When he added a diamond instead of coal to the molten iron, he again got steel. That experiment demonstrated that both diamonds and coal contain the same element. It was an expensive way to learn, but it worked.

Organizing Elements

About the same time, English chemist and all-round inquisitive person John Dalton studied two gases for which he had no name. He observed that one of them, which we now know was carbon monoxide (CO), contained, by weight, four parts oxygen to three parts carbon. The other gas, which we now know was carbon dioxide (CO_2), had, by weight, eight parts oxygen to three parts carbon. He interpreted this to mean that CO_2 contained twice as much oxygen as CO did.

Dalton studied the weights he obtained and decided that a carbon atom must weigh only three-fourths as much as an oxygen atom. From that point, he was able to propose an entire system of classifying chemical elements based on the differences in their atomic weights. He explained his reasoning in a book called *New System of Chemical Philosophy*. In that book, he dealt the final death blow to alchemy by explaining that an atom of one weight could not be changed into an atom of another weight by chemical means. The long-held dream of changing common elements into gold was over.

Actually, Dalton was not too meticulous in his measurements, and though his idea was correct, his results were often inaccurate. It took Jöns Jakob Berzelius, a Swedish physician who was more interested in chemistry than medicine, to correct Dalton's figures. He measured thousands of compounds, relating them all to a base weight of 1 for hydrogen (H, element #1). The weights he determined were amazingly close to those reached later by more refined techniques.

The work of Dalton and Berzelius prepared the ground for Dmitri Mendeleev. He was the Russian chemist who developed the Periodic Table of the Elements as we now know it.

John Dalton

Element Number 6

On the Periodic Table of Elements (pages 4 and 5), carbon is element number 6, and is given the symbol C. It is the first element in Group 14—sometimes called 4A. All the members of that group have only four electrons in their outer, or valence, shell. A full outer shell would have eight electrons. Because carbon's valence shell is not full, the element tends to react easily with other elements. In chemical reactions, individual carbon atoms may either lose four electrons, gain four, or share four. They are most likely to share four electrons, which is why carbon can combine with so many other elements. The other similar tetravalent (having four valence electrons) elements in Group 14 are silicon (Si, element #14), germanium (Ge, #32), tin (Sn, #50), and lead (Pb, #82).

As element number 6, carbon usually has six neutrons and six protons in its nucleus, as well as six electrons in orbit around the nucleus—two in its inner shell and four in its valence shell. But there are other forms, or isotopes, of carbon that have a different number of neutrons in their nuclei. Isotope means "same place." This name was used because several different isotopes of an element occupy the same place in the Periodic Table. Each isotope of carbon is still element number 6, because the identification of an element depends on the number of protons, not neutrons, in the element's nucleus. However, the isotopes of carbon have different mass numbers than the usual carbon atoms because of the additional neutrons in their nuclei. The six neutrons and six protons of the most common forms of carbon atom give it a mass number of 12. It is called carbon-12.

For almost two centuries after John Dalton did his work with organizing elements, the Periodic Table was based on hydrogen having an atomic weight of 1. The weights of all other elements were related to that fact. However, in 1961, scientists agreed that they would, from then on, change the comparison standard from hydrogen to carbon-12.

Although carbon-12 is the most common carbon isotope, atoms of carbon-13 are also naturally occurring isotopes. Carbon-13 also has six protons, but it has seven neutrons. Chemists assume that all samples of carbon will have the isotopes mixed in the same proportions, and so carbon in general is assigned an atomic weight of 12.011.

Reacting Together

Even though carbon joins up readily with other elements to form compounds, the reactions usually don't occur at room temperature. Fluorine (F, element #9) is the only element that reacts at room temperature with carbon, and even that reaction takes place rather slowly. But if carbon is heated, reactions occur quickly. When carbon combines rapidly with oxygen, the result is combustion, or fire.

Carbon also reacts with most of the elements called halogens. Found in Group 17 (sometimes called 7A) of the Periodic Table, halogens include fluorine, chlorine (Cl, element #17), and bromine (Br, element #35). All these elements have only seven electrons in their valence shell and readily link up with carbon to fill the eighth spot.

Carbon does not melt to a liquid when it is heated. At a temperature of about 3,550°C (6,420°F), carbon changes directly from a solid to a gas. This process is called sublimation, and it is the same one that changes solid carbon dioxide—better known as dry ice—directly into a gas.

The dark layer, or stratum, in this rock is a seam of coal.

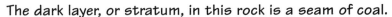

Carbon-bearing globules (the orange spheres shown above) in a meteorite that is believed to have come from Mars may indicate that there were once living things on that planet.

Because of its structure, carbon forms more compounds than almost all the other elements put together—probably at least a million. This makes carbon one of the most widely distributed elements in our physical world. It is a prime ingredient in calcium carbonate (the rock called limestone), which was formed from the many billions of living things that inhabited ancient seas. (Calcium is Ca, element #20.) Coal is mostly carbon, and petroleum and natural gas—also formed in ancient times—are chiefly hydrocarbons—compounds containing only carbon and hydrogen. Coal, petroleum, and natural gas are all fossil fuels, formed in ancient times from the remains of plants and animals.

Carbon is also found in our atmosphere, as carbon dioxide and carbon monoxide. Plant life depends on taking in CO_2 to survive and grow. Animal life, in turn, depends on taking in and digesting plants and other animals, thus getting the carbon they need second-hand.

Carbon occurs in large quantities in our solar system and throughout the universe. The atmosphere of Mars consists of at least 96 percent carbon dioxide. Stars also contain carbon. In general, cooler stars contain more carbon than hotter stars. These cooler stars are usually reddish in color.

THE
FOUR FACES
OF CARBON

To some people, diamonds mean romance, marriage, and beauty, but the sparkling gemstones we know as diamonds are actually a form of carbon that has, at some time in the ancient past, been subjected to great heat and pressure. The heat and pressure formed the hardest natural mineral known. It is so hard that the only thing that can cut a diamond is another diamond. Diamonds not used in jewelry are used by industry as points on cutting tools and as an abrasive to polish or grind other materials.

Diamonds are only one of the four forms, or allotropes, of carbon. The other three forms are graphite, amorphous ("without shape") carbon, sometimes called soot, and the recently discovered form called fullerenes. These

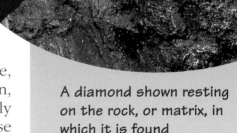

A diamond shown resting on the rock, or matrix, in which it is found

forms occur because the structures into which the carbon atoms are arranged are different in each allotrope.

The atoms in diamonds and graphite are held together in covalent bonds, meaning that they share pairs of electrons. Covalent bonds are very strong. Diamonds have their carbon atoms arranged in hexagons (six-sided shapes) that repeat themselves in fixed three-dimensional patterns. This three-dimensional structure makes a very strong crystal lattice.

On the Mohs Hardness Scale, diamond is at the top—number 10—the hardest natural mineral known. Diamonds can scratch all other materials as well as other diamonds. Graphite is so different from diamond that it drops down to the bottom of the Mohs Hardness Scale, having only a 1 or 2 rating—ranking graphite among the softest minerals known.

Slippery Graphite

The only time most of us come into contact with graphite is when we write with a "lead" pencil. The so-called lead is a mixture of graphite and clay molded into long, thin rods and then wrapped in wood. The name graphite is based on the Greek word *graphein,* meaning "to write".

Like diamonds, graphite has its atoms arranged in hexagons, but the layers of hexagons don't link up three-dimensionally. Instead, the layers of carbon molecules are free to slide across one another. This is what makes graphite so soft, and slippery enough to be used as a lubricant in heavy machinery. Graphite also leaves a layer of itself behind when it is rubbed on paper, as a pencil does. Because graphite is not affected by great heat, it is mixed with clay to make the bowl-like containers, or crucibles, in which other chemicals are heated to high temperatures.

Graphite is often found naturally in both large sheets and as small round lumps within other rocks. The rocks were formed by heat and pressure. If the pressure had been a bit stronger or had

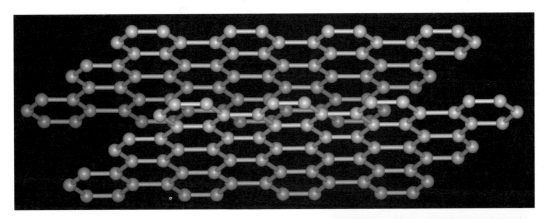

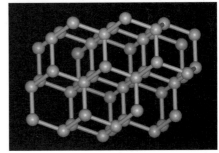

The two forms of carbon shown here have six atoms arranged in hexagons (six-sided shapes). In diamonds (right), the hexagons are formed vertically as well as horizontally, making strong crystals. In graphite (above), the hexagons form layers that are only slightly attracted to each other and can slide across one another.

lasted a bit longer, those graphite lumps might have turned into diamonds.

Most industrial graphite is produced from coke in a process developed by Edward Acheson. A chemist who had earlier worked for Thomas Edison, Acheson was trying to create synthetic diamonds when he produced graphite in 1896.

It was another fifty years before the first synthetic diamond was made from graphite. The General Electric Company put graphite under the pressure of 50,000 atmospheres, or 730,000 pounds of pressure per square inch, at a temperature of 3,000°C (5,400°F) for sixteen hours to produce the first artificial diamond. Today, synthetic diamonds are produced at somewhat lower temperatures but under much greater pressures.

In the first controlled nuclear chain reaction, carried out at the University of Chicago in 1942, graphite rods were inserted into a nuclear "pile" or reactor to slow down the speed at which nuclear particles moved. This is a painting of the historical scene.

Graphite has been used in nuclear reactors since the first controlled nuclear chain reaction was carried out at the University of Chicago in 1942. Rods of the carbon material are inserted into a reactor to slow down, or moderate, the speed at which neutrons crash into other atoms, making nuclear reactions take place. If the neutrons weren't slowed down, the result would be a nuclear explosion instead of a controlled nuclear reaction.

Carbon Without Shape

Amorphous carbon, the third form of carbon, is the black powder left after a carbonaceous (carbon-containing) object is burned. One kind of amorphous carbon is soot, which is basically unburned carbon. Soot is also known as carbon black and lampblack. If you've seen a flame make a black smudge on the glass chimney of a lantern, you've seen carbon black.

Unlike diamonds and graphite, the atoms in amorphous carbon have little attraction for each other. Some graphite is more like diamond in being very structured, and some is more like amorphous carbon, in having little structure. The atoms of coke being heated in a furnace along with ash acquire a more regular structure as the temperature rises. The carbon in charcoal from wood, on the other hand, never acquires a regular structure regardless of how high the temperature gets.

Soot seems as if it should be just a waste product, but it is amazingly useful. It has long been an ingredient in printing ink, for example. The soot used in industry is produced by burning methane gas (CH_4) with limited amounts of oxygen. The soot left behind is scraped off the walls of the burning chamber.

More than two-thirds of the "rubber" in a rubber tire is actually carbon black. Without it, the rubber would disintegrate when it got hot. Pollution in the air also harms rubber that does not contain soot.

The "Soccer Ball" Accident

In the late 1980s, British and American scientists working with Harry Kroto and Richard Smalley at Rice University in Texas were studying a plasma, a gas consisting of electrically charged particles. Plasma is sometimes called the fourth state of matter (after solids, liquids, and gases). Although plasmas exist naturally, such as in the upper atmosphere, many are now created in laboratories. The scientists in Texas were trying to find out if amino acids, the basic molecules of life, could be created out of gases in the universe.

Kroto, Smalley, and their colleagues were experimenting with a "brew" of carbon atoms and ions (atoms with an electron added or removed). They shot a laser beam through the brew, which stirred and churned. When they analyzed the results, the scientists were astonished to find out that the carbon atoms in the brew had joined together to form perfect spheres that looked like soccer balls.

Buckminster Fuller's geodesic dome was built in Montreal, Quebec, Canada, for the Expo '67 World's Fair.

Eventually, the peculiar molecule discovered in the lab would be named buckminsterfullerene, after American architect Buckminster Fuller. He designed the geodesic dome, which consists of many triangular-shaped sides. The scientists thought that their carbon molecule looked like the huge dome Fuller had built in Montreal. The buckminsterfullerene molecule is described by mathematicians as a truncated icosahedron, but it became popularly known as a "buckyball."

Like a real soccer ball, the buckyball—also called carbon-60—has sixty points at which five-sided and six-sided structures join. The structure has twelve pentagons, or five-sided shapes. Each pentagon is surrounded by five hexagons, or six-sided shapes, for a total of twenty hexagons. Each of the atoms is linked to three others, two by single bonds (using two of carbon's valence electrons), and one by a double bond (using the other two).

The initial discovery of the buckyball was a pleasant accident, but it was not easy proving to others that it existed. The discoverers at Rice University built a device that evaporated graphite and produced ten grams (0.3530 oz) of soot. About one-tenth of the soot was buckminsterfullerene. By 1990, the new molecules were being produced in commercial amounts and became available for scientists anywhere to study.

This model of the "buckyball"—carbon-60—shows that each carbon atom (located at the points in the six-sided and five-sided rings) is attached to three other carbon atoms.

Useful Fullerenes

Some scientists objected to carbon-60 being regarded as a real allotrope of carbon. They called the material fullerite. Within a few years, several physicists discovered that they could deliberately make such molecules. All fullerenes have an even number of atoms in their molecules because each time a hexagon is added, it takes two carbon atoms to fill it. A giant fullerene has been made containing 540 atoms.

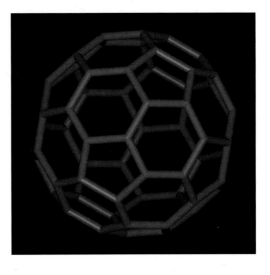

As fullerenes were being explored, it was also found that graphite could be evaporated and made to form thin, tubelike "whiskers." These "whiskers" are often called buckytubes because the ends of the tubes are spherical, like buckyballs. Buckytubes are amazingly strong, perhaps as much as 600 times stronger than steel. They may be very useful in microcircuits to carry electricity. Other scientists have made bucky "donuts," which are buckytubes with their ends joined together to form donut-like circles.

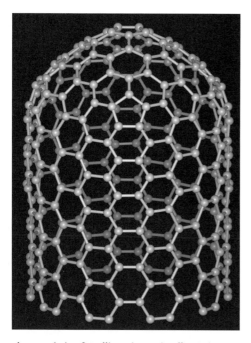

A model of a "buckytube," with carbon hexagons forming a tube that can enclose other atoms and molecules

A scientist at Georgia Institute of Technology has carried out a chemical reaction inside a buckytube, making it the world's smallest test tube. He inserted molecules of silver nitrate ($AgNO_3$) (silver is Ag, element #47; nitrogen is N, #7) into a buckytube only 0.00000508 mm (0.0000002 in) in diameter, and then bombarded the molecules with electrons. The bombardment caused the silver nitrate to heat, releasing the nitrate and leaving pure silver behind.

Some Old Buckyballs

It turns out that buckyballs are not just a novelty created in laboratories. Almost 2 billion years ago, a mountain-sized meteorite struck the earth in Canada.

Jeffrey Bada of Scripps Institution of Oceanography investigated some rocks from the crater the Canadian meteorite made when it struck our planet. He expected that all carbon would have disappeared long ago, probably burned up when the meteor entered Earth's atmosphere or eaten by bacteria over the years.

Instead, Bada found some buckminsterfullerene molecules in the rocks. A few of the cage-like structures contained helium atoms. Helium was one of the original elements in the universe. Bada's colleagues were certain that the buckyballs had to have been formed in outer space, perhaps beyond the solar system, among old, red giant stars.

CARBON PLUS ONE OR TWO

In another experiment, Jan Baptista van Helmont called the invisible gas given off by burning charcoal *air sylvestre* ("air from wood"). We now know the gas was also carbon dioxide. The air we breathe contains about 0.03 percent carbon dioxide. If the air contained much more than that, it would not be healthy for us.

In the late 1700s, English scientist Joseph Priestley noted that he had trouble breathing in a room where many candles had been burned. He enclosed some of that room's air in a glass chamber and found that a mouse would become unconscious if it were enclosed in the chamber. It is now known that if a person remains in a chamber with air containing as little as 5 percent CO_2, the person will start panting, trying to get enough oxygen into his body. More than 5 percent can lead to unconsciousness, even death.

Joseph Priestley

In 1771, Priestley also found that he could put plants in the same room where many candles had been burned and, within a few days, the air would be fresh again. He tried using several different types of plants, in case this air restoration was a trait of a certain kind of plant. The air became breathable with any plant he used. Priestley concluded that plants, in general, must give off something that was good for living animals. That something is oxygen (although it was still not named).

Dutch scientist Jan Ingenhousz took up Priestley's work and discovered that plants get the carbon they need from the air, instead of from soil, as was long thought. He also demonstrated that plants take in the air containing that element only during the day. At night, they take in oxygen, just like animals.

In 1787, French scientist Antoine Lavoisier published a book in which he described a plan for naming chemicals based on the elements in them. For the first time, Helmont's "air sylvestre" acquired a scientific name, carbon dioxide.

Dry Ice and the Rare Liquid

Carbon dioxide occurs naturally as a gas. It can be compressed into a liquid, but it remains a liquid only if kept under a pressure about five times normal atmospheric pressure. Called supercritical CO_2, the liquid's main use is in decaffeinating coffee. It is used to soak the green, unroasted coffee beans, from which it dissolves caffeine, the chemical that makes some people jittery after drinking coffee.

Used more often is CO_2's solid form—dry ice. As the gas is chilled, it becomes solid at −78°C (−109°F), without a liquid stage in between. When warming, dry ice sublimes (turns directly into gas), leaving no liquid to dispose of. That trait makes it very useful in refrigerating items for shipping. Sometimes the producers of a theatrical play will use dry ice to create the effect of fog on stage. A fan blows across containers of dry ice placed

Dry ice being used in a refrigeration unit to store medicines. Frozen carbon dioxide keeps things cold as regular ice does, but dry ice does not turn into liquid as it thaws.

backstage. Visible white vapor forms as the evaporating dry ice chills water vapor into droplets. The white vapor is blown onto the stage, where it lingers near the floor, looking spooky.

A Medal for a Fizzy Drink

When Joseph Priestley was studying the gas that was later called carbon dioxide, he could obtain all he wanted of it because he lived near a brewery where beer was made. One of the processes of beer-making is fermentation, which turns grains into alcohol, giving off CO_2 as a by-product which can be collected. In 1767, Priestley dissolved some of the gas in water and took a taste of it. Usually it is dangerous to taste a chemical experiment, but, in this case, Priestley discovered that the fizzing water had a pleasantly tangy taste. The Royal Society gave Priestley a medal to honor his discovery of carbonated water.

If you look at a sealed bottle of soda, you don't see any fizz. The CO_2 is dissolved in the liquid under pressure. But the instant you open the bottle, the pressure is released. The gas separates

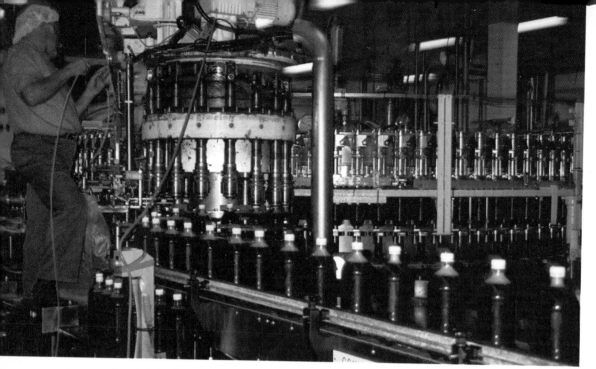

In a soft-drink plant, carbon dioxide is injected into the bottles just before they are sealed. There is no fizz until the containers are opened again.

out as bubbles, which rise to the top, with a fizzing noise. If the bottle or can is left open, the remaining gas bubbles escape and the leftover drink is "flat."

At first, most people thought that the carbonated water was good only as a medicine. Coca-Cola®, which was invented in the late 1800s, was originally sold as a tonic.

The Smothering Gas

Carbon dioxide is not, in itself, poisonous, but if it prevents a person from getting oxygen, it can be deadly. One night in August 1986, death came to many people who lived along the shores of Lake Nios in Cameroon, Africa. Apparently a huge bubble of CO_2, given off by the molten interior of the Earth, collected just beneath the sediments in the bottom of the lake. A land tremor or other event weakened the upper surface of the bubble and it exploded, sending a cloud of smothering CO_2 across the villages. With no warning, more than a thousand people died.

Such an event was tragic, but the same ability of carbon dioxide to shut out oxygen can also save lives. Most home fire extinguishers contain compressed carbon dioxide. When the valve on the fire extinguisher is pressed, the released gas expands rapidly, forming a layer that is heavier than air, so it blankets the fire. Even though CO_2 contains oxygen, the oxygen in CO_2 will not support combustion, so the flames are quickly extinguished.

The Dinosaur Killer

No one knows for sure—and we may never know—just why dinosaurs became extinct. In one widely accepted theory, an asteroid entered Earth's atmosphere about 65 million years ago. Moving at more than 80,000 km (50,000 miles) per hour and glowing red-hot, it struck the planet's surface, melting rock and vaporizing all living things within a huge area. Fires flared all over the planet, spewing so much ash into the atmosphere that the sky was darkened and no sunlight could penetrate it. Ash from the fires coated those plants that remained, killing them.

Those animals that survived this holocaust continued to exhale carbon dioxide. But few plants remained to give off oxygen, and soon there was little oxygen left for the dinosaurs to breathe. Their own exhalations of carbon dioxide began to poison the atmosphere.

The theory holds that some animals and plants survived this terrible time of choking darkness, and Earth's own healing powers gradually cleared the skies. The survivors began to repopulate the planet and to bring the atmosphere into balance. But it was too late for the dinosaurs that continue to fascinate us.

A scientist releasing a balloon attached to an instrument that measures carbon dioxide in the air. The small instrument will radio measurements back to the ground.

Global Warming

Carbon dioxide may be building up in the atmosphere again, though much more slowly than it did about 65 million years ago. During the twentieth century, the amount of carbon dioxide in the atmosphere has increased almost 15 percent.

The sun gives off many invisible rays, or waves, of energy, in addition to visible light. Some of the rays, called ultraviolet rays, hit the planet's surface, change into heat waves, and are reflected back into space. These heat waves can be absorbed by certain gaseous molecules in the atmosphere. The heat is held in the atmosphere, warming the planet, much like the sun's rays warm the air in a greenhouse. Without these "greenhouse" gases, Earth's surface would always be about −17.8°C (0°F), which is too cold for survival of most living things.

The molecules of greenhouse gases consist of three or more atoms. CO_2 is the primary greenhouse gas. Others include methane (CH_4) and nitrogen dioxide (NO_2). Actually, plain old water vapor (H_2O) is also a greenhouse gas. But it's the increase in the amount of the other gases, especially CO_2, that makes them potentially dangerous.

Many scientists believe that the human use of fossil fuels such as burning coal in factories and using gasoline in millions of automobile engines is increasing the amount of carbon dioxide in the atmosphere. As a consequence, they think that the planet is slowly getting warmer. If the trend continues, the oceans might rise, flooding low-lying lands throughout the world, because warm water takes up more room than cold water.

Other scientists disagree. But we can't wait for millions of people to lose their homes, cities, and agricultural areas to find out who's right. Many nations have agreed that we must cut CO_2 emissions. The United States is one of the world's major producers of atmospheric CO_2 from factories that burn coal and internal combustion engines that burn petroleum products.

Without that Extra Oxygen Atom

Oxidation can occur within a cell or a forest. No matter where it occurs, if it involves living, or once living, material, one result of oxidation is the formation of carbon dioxide:

$$C + O_2 \rightarrow CO_2$$

Internal combustion engines aren't as efficient as they could be in oxidizing the carbon in gasoline. Instead of forming all CO_2, internal combustion engines also form some CO, or carbon monoxide. Carbon monoxide is much more dangerous to humans than carbon dioxide.

Everyone is exposed to carbon monoxide at some time. About one-millionth of the atmosphere is naturally CO, and automobile engines usually give off as much as 10 percent CO. But if the exposure is greater, trouble can arise.

The United States government has set a standard of an average exposure for humans of 9 parts CO per million (ppm) parts of air during an eight-hour day as a danger level, or 35 ppm exposure for one hour. It is probably safe for a driver to sit in a heavy traffic jam for up to an hour, but it would be dangerous if the exposure were to continue. People who live near a traffic interchange in a city—where there might be an almost constant traffic jam—could suffer from continual exposure to carbon monoxide. Both their minds and bodies could be affected.

Carbon monoxide has no smell, so people may not know when it is in their homes. Sometimes people die if a heater, for example, functions incorrectly and gives off CO. For that reason, people should install carbon monoxide detectors in their homes.

Unlike smoke detectors, which sound an alarm whenever smoke is detected, carbon monoxide detectors measure the CO in the air over a period of time. Residential CO detectors are usually set for the alarm to sound when 100 parts CO per million parts of air are detected for at least 90 minutes.

Carbon monoxide is useful in many industrial processes because of the ease with which the gas reduces (or takes oxygen away from) other chemicals. Reduction with carbon monoxide is commonly used to turn an ore from a mineral oxide into a pure metal. Iron ore, for example, is turned into pure iron by blasting it with carbon monoxide. The CO in this mineral becomes CO_2 when treated this way.

Endangering Life

In the human body, oxygen is taken in through the lungs. Deep in the alveoli, the tiny chambers in the lungs, oxygen moves through the lung walls into the capillaries, the smallest blood vessels. Once in the blood, the oxygen molecules become part of the red blood cells, where they link up with a complex molecule called hemoglobin.

Probably more is known about hemoglobin than about any of the other complex molecules in our bodies. This amazing molecule has the chemical formula:

$$C_{2954}H_{4516}N_{780}O_{806}S_{12}F_4$$

A simplified model of the complex hemoglobin molecule in blood. If an oxygen atom attaches to the molecule, it supports life. If carbon monoxide takes the place of the oxygen, the CO can bring death.

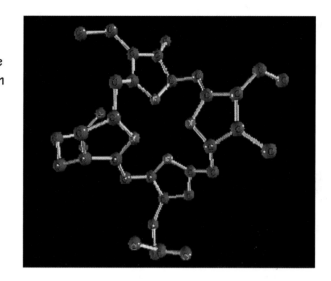

Oxygen attaches itself to the hemoglobin molecule to travel to the tissues. It then separates from the hemoglobin, enters the cells, and oxidizes other chemicals, especially sugars, releasing energy. A product of that cellular combustion—just as in a forest fire—is carbon dioxide.

Some carbon dioxide released by the cells is once again attached to hemoglobin and travels back to the lungs. But most of the carbon dioxide actually becomes part of the blood fluid. Either way, it gets back to the lungs where it is exchanged during breathing for a new supply of oxygen.

But the process is not the same if we inhale carbon monoxide. The CO molecule does indeed attach itself to the hemoglobin, but it doesn't detach itself at the tissues. Instead, it clings to the hemoglobin, preventing oxygen from attaching to this vital molecule. With alarming speed, the CO takes over enough hemoglobin so that oxygen cannot reach the body's organs, especially the brain. The result can be death.

CARBON IN LIVING THINGS

Human beings and all the other life on Earth make up what is called carbon life. Scientists think that the carbon in our known universe developed in red giant stars. Eventually, some of the whirling debris formed our solar system, with Earth as the third planet from the sun.

An important element in the early atmosphere and oceans of Earth was carbon. The theory is that as the molecular brews formed and re-formed on Earth, some complex molecules took on the ability to duplicate themselves. Life on Earth had begun, and it was carbon based.

The Carbon Cycle

Carbon moves through our world in a cycle that involves all living creatures, as well as many nonliving things. Since the carbon cycle is a full cycle, there is no specific starting or ending point to it.

In ancient times, carbon from dense forests of living plants and animals laid down the deposits that formed coal in our Earth.

32

However, we'll begin our look at the carbon cycle with the carbon found in air.

The carbon in air is in the form of carbon dioxide and a little carbon monoxide. Only about 0.03 percent of the air making up the lower atmosphere of Earth is carbon dioxide. Carbon monoxide appears naturally only in very small traces.

Carbon dioxide is taken into plants through openings called stomata. As sunlight strikes the leaves, the CO_2 combines with water and forms sugars. The sugars, particularly the simplest one called glucose, are utilized by the plant's cells as a source of energy to grow and reproduce.

Simply put, this process—called photosynthesis—is the following chemical reaction:

carbon dioxide + water + sunlight = glucose + oxygen

$$6CO_2 + 6H_2O + \text{sunlight} \rightarrow C_6H_{12}O_6 + 6O_2$$

The oxygen resulting from photosynthesis is released back into the air by the plants. Animals breathe in that oxygen and use it to burn, or oxidize, the sugars gained from eating plants and animals. The oxidation process results in energy for the body and carbon dioxide. The CO_2 is exhaled into the air, where it is available for plants to use again.

However, like animals, plants need oxygen to oxidize the sugars they make. As usual with oxidation, the result is carbon dioxide. Plants take in carbon dioxide during the day, giving off oxygen. At night, the same process as animal respiration occurs—the plants take in oxygen, burn sugar for energy, and give off carbon dioxide.

Fueling the Body

The energy needed by all living things to grow, reproduce, and repair themselves comes from carbohydrates. These are the many molecules that contain carbon, hydrogen, and oxygen.

The basic carbohydrate unit is a saccharide. A saccharide unit is not broken down further until it is oxidized, or burned, for energy. The type of sugar consisting only of individual saccharide units is called a simple sugar, or monosaccharide. The monosaccharide glucose is often called blood sugar because it is the form in which sugar moves through the blood and enters the cells. All carbohydrates eventually break down into glucose. Pure glucose in water is fed to patients in hospitals through needles going into veins in their arms. This glucose solution keeps their energy level up when they can't digest regular food or fluids.

Fructose, the sugar in most fruits, is also a monosaccharide, though it is much sweeter than glucose. Glucose and fructose have exactly the same chemical formula—$C_6H_{12}O_6$—but the atoms are arranged differently.

Sucrose is common table sugar (and the most common chemical produced by industry). Its chemical formula is $C_{12}H_{22}O_{11}$. Sucrose is a disaccharide (two-unit sugar) made up of joined molecules of glucose and fructose. Table sugar is made by melting the sugars in sugar cane or sugar beets and letting it evaporate into crystals. Lactose, which is the sugar in milk, is also a disaccharide. It breaks down into two monosaccharides—glucose and galactose—for use by the body.

Complex carbohydrates, called polysaccharides, may be made up of thousands of saccharide units. A molecule of starch, for example, contains thousands of glucose molecules joined end to end. Starch is the way carbohydrates are stored by plants such as beans, corn, potatoes, and wheat.

Carbohydrates are used as fuel by humans. As they are eaten and digested, carbohydrates break down into complex sugars and enter the bloodstream, which carries them to the liver. The final conversion to usable sugars takes place in the liver. Then these little bundles of energy are sent to all parts of the body.

Bread dough is quite thick and heavy, but yeast in the dough produces bubbles of carbon dioxide that make the dough light and airy. The loaf of bread in the background has already risen from the yeast's action.

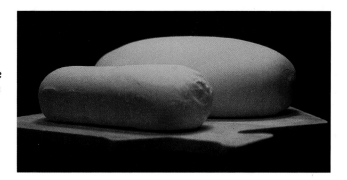

Bread is the most widely eaten food in the world and can be found in various forms everywhere. White bread is the most popular bread in the United States. White bread is made from flour that has had the outer hull of the wheat grain removed. White bread consists of 50 percent carbohydrates, 36 percent water, 9 percent protein, 3 percent fat, and 2 percent ash.

Bread dough rises (increases in size and fluffiness) by the same fermentation process that makes alcohol in beer. To make bread, flour is mixed with a liquid and yeast. Chemicals in the yeast react with sugar, producing CO_2. In beer, the CO_2 is given off. In bread, it makes the dough bubble up.

In animals, carbohydrates are stored in the liver and muscles as a chemical called glycogen. As the body needs sugar in the blood, the glycogen is broken down into simple glucose. Gerty and Carl Cori, who described this process in detail, won the 1947 Nobel Prize for Physiology or Medicine for their work.

It has been recommended that nearly 60 percent of our diet should consist of carbohydrates. No more than 10 percent of that amount should be made up of processed sugars such as those found in candy, cake, and pie. That means we need lots more unprocessed grains, vegetables, and fresh fruits in our diets to provide the carbohydrates we need for healthy living. Without enough natural carbohydrates, muscle structure, fluid balance, and metabolism can suffer.

Uncleaned cellulose fluff, just as it has been picked from the cotton plant, is shown above. Cellulose is a complex carbohydrate.

The Tough Part of Being Green

The fluid in plants is a mixture of several simple sugars—especially glucose and fructose. They are held in by another "ose"—cellulose, a complex carbohydrate containing thousands of glucose units. This repetition of the same glucose unit over and over is called polymerization. Because cellulose is a polymer, it cannot be broken down by the human digestive system. Only animals that chew a cud (partially digested food that is chewed again) such as sheep and cows, can digest grass or most other plant cellulose. It's not actually the cow's or sheep's stomach that is digesting the grass. It is the microorganisms in their digestive tracts that break the cellulose down into usable chemicals.

But people need cellulose in their diets. We call it fiber. We don't digest it, but the cellulose keeps our digestive systems working properly. Most fruits, as well as the bran, or outer hull, of whole grains supply us with fiber.

The purest cellulose available naturally is from the puffy white fruit, or boll, of the cotton plant. The fiber inside the boll consists of up to 90 percent cellulose.

Like cellulose, starches are polymers of glucose. In fact, cellulose and starch have the same chemical formula, but the way the atoms are arranged is different. Corn, wheat, potatoes, and other important plants in our diet contain both starch and cellulose. The digestive enzymes in our small intestines can break starch down into its glucose molecules but not cellulose.

Fats and Oil

We need a regular supply of carbohydrates every day because they are used immediately for energy. A different category of nutrients, called lipids, can be stored by the body for later use, much to the dismay of overweight people. Lipids are greasy substances that will not dissolve in water. They are primarily fats.

Some fats are solid or semi-solid at room temperature. These fats tend to come from animal sources. Other fats are oils, which are liquid at room temperature. They tend to come from plants.

Fats are stored in tissue called adipose tissue, which cushions the organs in our bodies, keeping them both warm and protected from blows. Fats also make up a considerable portion of the cell membrane of every cell.

The main components of fats are fatty acids. There are about forty different fatty acids. They differ primarily in how many carbon atoms they have. Many of them are made by our bodies from other fatty acids. Linoleic acid and linolenic acid are the only two fatty acids we need that our bodies can't manufacture. Most common plant oils, especially soybean oil, are high in linoleic acid. Linolenic acid is found especially in fish oils.

Fats and Hearts

You've probably heard the terms *saturated, unsaturated,* and *polyunsaturated* in reference to fats. These terms describe the carbon atoms in the fatty acid chains. The carbon atoms in a

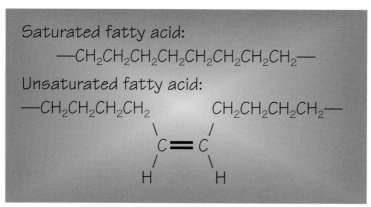

Saturated fatty acid:

$$-CH_2CH_2CH_2CH_2CH_2CH_2CH_2CH_2-$$

Unsaturated fatty acid:

$$-CH_2CH_2CH_2CH_2 \qquad CH_2CH_2CH_2CH_2-$$

Each carbon atom in a saturated fatty acid is attach to as many hydrogen atoms as it can be. An unsaturated fatty acid is missing some hydrogen atoms.

saturated fat have their valence shells completely filled (saturated) by electrons from hydrogen atoms or other carbon atoms.

An unsaturated fatty acid has at least two carbon atoms joined by a double bond and so they are attached to fewer hydrogen atoms. The two carbon atoms share four electrons. A polyunsaturated fatty acid has several such carbon-carbon bonds, leaving room for more atoms to join up. They are not completely filled.

Fats that are unsaturated—those that can have other atoms attached to them—seem to be healthier for us. The chemical called cholesterol, for example, can probably be prevented from harming our hearts if we avoid eating too many saturated fats.

Cholesterol

Cholesterol makes up part of cell membranes. It controls the way various fluids move in and out of the cell. It also is the basic chemical from which various biological chemicals called steroids, such as sex hormones and bile, are made. Cholesterol, in general, is a good thing, a vital thing, in our bodies. We couldn't function without it. In fact, about 10 percent of our brain consists of cholesterol.

Unfortunately, in some people, cholesterol has a tendency to accumulate as part of a yellowish deposit called plaque that may build up inside blood vessels, restricting the way blood flows through them. If the build-up of plaque occurs near the heart, the blood vessel can become blocked, causing a heart attack.

The good news is that the kinds of fats we eat—saturated or unsaturated—can affect the build-up of cholesterol in the blood vessels. This happens through chemicals called lipoproteins, which carry cholesterol throughout the body.

One type of lipoprotein called low-density lipoprotein, or LDL, may deposit cholesterol in the arteries. Another type, high-density lipoprotein, or HDL, carries cholesterol to the liver, from which it can be excreted in urine. Therefore, we want more HDL than LDL because HDL can actually remove cholesterol from artery walls. Eating unsaturated fats—as well as getting plenty of exercise—are associated with the production of HDL. Most of us naturally have more LDL in our systems, and the body easily makes more when our diet includes too many saturated fats.

The Many Faces of Protein

The nutrients called proteins differ from carbohydrates and fats by having nitrogen in their chemical makeup, in addition to carbon. Chemists of 100 years ago called the complex chemicals that make our bodies function proteins, meaning "first." They thought proteins were the most important organic chemicals.

Protein molecules in any living thing differ according to what plant or animal they are in, and even according to where in the living body they are. In humans, for example, hair, blood, and heart muscle are each made up of many different kinds of proteins. There may be well over 10 billion different proteins in existence. Amazingly, all those proteins are constructed of only twenty different kinds of molecules called amino acids.

All amino acids have the same primary structure, like that shown on the next page, along with a model of glycine, the

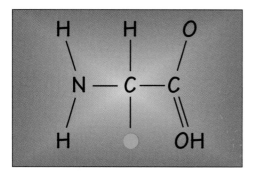

The green dot in the structural formula above represents the spot where different types of molecules attach to the amino acid group to make various amino acids. The simplest amino acid in the human body is glycine (right). It has a single hydrogen atom attached where the dot is in the formula above. The red spheres in the model represent oxygen atoms.

simplest amino acid. A more complex one called arginine (the names of most amino acids end in "ine") has a longer chain of atoms attached in a straight line, of which this is just part:

$$\text{—CH}_2\text{—CH}_2\text{—CH}_2\text{—NH—C(NH)NH}_2$$

The way these amino acids are put together in different arrangements and sizes determines which type of protein is being made. Two proteins may be made up of exactly the same amino acids, but if one has the amino acids strung out in a long chain, its function is probably quite different from a protein with a short chain of amino acids.

Proteins are used by every cell in the body, but it takes many different kinds of proteins to properly nourish the whole body. Meat contains the complete set of amino acids our bodies need. A vegetarian (a person who doesn't eat meat) can get a complete set of amino acids by combining several different vegetables in one meal.

After all the protein needed by the body for its normal tissue repair and growth is used, the protein left over is treated like a carbohydrate and used for energy. The nitrogen is removed and excreted in urine.

Carbon in the Ring of Life

Probably the most significant chemical that pops up in the news regularly these days is deoxyribonucleic acid (DNA). Molecules of this very complex acid make up human chromosomes, which contain the genetic material that is passed on from one generation to the next. DNA molecules have the amazing ability to split, so that only half their genetic information goes into each female egg or male sperm. When an egg and sperm unite, a new individual, with genetic information from two parents, begins to develop. That development takes place as instructed by the DNA.

Amazingly, the DNA molecules that contain enough information to form billions of different humans are made up of only four kinds of molecules called nucleotides. The four nucleotides are called adenine, guanine, cytosine, and thymine. They all have at their center a six-sided ring consisting of carbon and nitrogen.

DNA is often mentioned in news stories for several reasons. In recent years, scientists have become able to analyze an individual's DNA makeup from hair or blood left at a crime scene. This DNA evidence helps determine whether that person—and, in all probability, no one else—might have committed a crime.

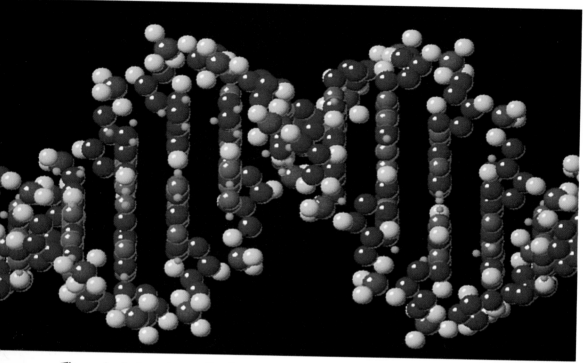

The spiral structure of a DNA molecule shows many smaller molecules, called nucleotides, connected to each other crosswise in the spiral. The four nucleotides in the human body can be arranged in enough different ways to create billions of genes.

In other DNA experiments, DNA scientists called genetic engineers are studying the entire structure of human chromosomes. They are working to identify every gene and what it does. They especially hope to locate the individual genes that control various inherited diseases. They hope to be able to replace the genes that cause the diseases with genes that will gradually heal the person.

THE CHEMISTRY OF CARBON

Early chemists were intrigued by the chemicals found in living things, or organisms. They were certain that such "organic" chemicals could be made only by living things. Then, in 1828, German chemist Friedrich Wöhler made in his laboratory the complex organic compound called urea, $CO(NH_2)_2$, which is found in the urine of almost all mammals. Wöhler's urea was an organic chemical that had not been made within a living body.

Well over 5 million organic compounds have been identified—both from natural sources and made in the laboratory. Many of these organic compounds are hydrocarbons, molecules made only of carbon and hydrogen.

A Pool of Hydrocarbons

Throughout the history of Earth, seas have covered huge portions of the land.

Methane, the lightest of all the hydrocarbons, being "burned off" at an oil refinery

43

Over millions of years, the plants and animals living in those seas died and sank to the bottom. As generation followed generation, the weight of those on top exerted great pressure on those underneath, and gradually changed the once-living materials into petroleum. The thick fluid oozed into pools between layers of rock and usually has stayed there until released by humans. Petroleum is the natural resource that provides a vast pool of hydrocarbons which we have put to use.

The same process that created petroleum also created natural gas. The main ingredient in natural gas is also the simplest of all hydrocarbons, methane (CH_4). Methane is used most often for heating and cooking.

Because the carbon atom has four electron spots available in its valence shell and because carbon cannot form a double bond with hydrogen, it takes the electrons from four hydrogen atoms to fill carbon's valence shell. The resulting methane molecule (CH_4) is as simple as a hydrocarbon can be. The structural formula for methane is shown at the top of the diagram on the left.

Actually, even the structural formula of methane still does not show the actual three-dimensional arrangement of the carbon and hydrogen atoms. The three-dimensional illustration next to the structural formula in the diagram shows the tetrahedral (four-faced angle) arrangement of atoms in methane.

The carbon and hydrogen atoms in methane are arranged as tetrahedrons or four-faced angles.

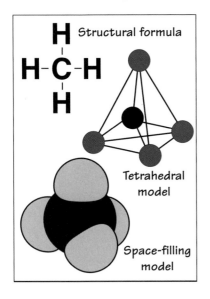

H Structural formula

H–C–H

H

Tetrahedral model

Space-filling model

Organizing the Organics

If one of the hydrogen atoms in methane is removed, a free radical, or part of a molecule, is left over. The free radical from methane readily reacts with other atoms to gain the eighth electron the carbon needs to have a complete outer (valence) shell. When two such free radicals join, they make ethane (C_2H_6) and each carbon atom is complete. After methane itself, ethane is the smallest of the group of hydrocarbons called alkanes.

Structurally, ethane looks like two attached methane molecules, with the central carbons linked, as shown at the right.

Other alkane molecules then get progressively larger. Each one is given a name with the "ane" ending. Propane is C_3H_8, butane is C_4H_{10}, pentane is C_5H_{12}, and decane is $C_{10}H_{22}$. These alkane gases are all colorless and odorless and burn easily in air. The higher-numbered alkanes are usually solid at room temperature.

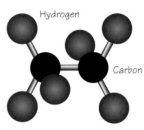

An ethane molecule

This pattern of increasing sizes is typical in organic chemistry. The study of organic compounds was chaotic until it was realized that the many compounds occur as series of related compounds. These series usually contain a similar group, or collection, of atoms arranged in a certain way.

Isomers and Structural Formulas

Hydrocarbons are significant because carbon has the ability to bond to itself and to form long chains of molecules, both straight and branched. The atoms also can arrange themselves structurally in different ways. Two chemicals with the same chemical formulas but with the atoms arranged differently are called isomers. This is one of the factors that makes hydrocarbon chemistry so complex and such a treasure trove of new chemicals for chemical engineers to develop. Alkanes, for example, can have numerous isomers. An alkane with 40 carbons and 82 hydrogen atoms has more than 62 trillion possible isomers.

The existence of isomers required chemists to come up with

formulas that showed more than the atoms making up a compound. The four carbon atoms in butane, for example, would be described in a standard chemical formula as C_4H_{10}. However, if the carbon atoms are arranged in a straight chain as shown by the dots below, the chemical formula would be $CH_3CH_2CH_2CH_3$.

• — • — • — •

If the carbon atoms are in a branching pattern, then the formula would read $(CH_3)_3CH$. In all three formulas there are still only four carbon atoms and ten hydrogen atoms.

• — • — •
|
•

If the groups of atoms in butane are arranged in two different ways, the resulting chemicals would have different characteristics. For this reason, chemists began to write hydrocarbon formulas in a way that shows their structures.

Other Hydrocarbon Families

Other families of hydrocarbons are both as logical and as complex as the alkanes. One family, called the alkenes, starts with ethylene (C_2H_4). The two carbons are held together by sharing four electrons, leaving two free spaces on each carbon outer shell where hydrogens can link up. The

The alkyne called acetylene is mixed with oxygen to create a gas that burns with a very hot flame that is useful in cutting metal.

same prefixes used with alkanes are used with the progressively larger alkenes. So we have propene (C_3H_6), butene (C_4H_8), and pentene (C_5H_{10}) at the beginning of the series. The alkenes are also called olefins.

Ethylene gas can be used to help fruit ripen. The gas is pumped into storage bins filled with unripe fruit, which then ripens. No harm is done to the fruit. In fact, its vitamin C content increases.

A third family is the alkynes. Alkynes have two carbon atoms that share six electrons, leaving only one electron to link to a hydrogen. The simplest alkyne is officially known as ethyne, C_2H_2, but it is generally known by its older name of acetylene. Acetylene is a colorless gas (with the smell of garlic) that burns with a very hot flame. It is especially useful when combined with oxygen for cutting metal. Acetylene is made by heating methane without oxygen.

The Benzene Ring

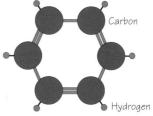

Carbon

Hydrogen

The benzene ring

In the early part of the 1800s, several scientists investigated whale oil and coal oil, which at that time were used as fuels for lighting. Residues from both fuels contained a chemical with the formula C_6H_6. In 1845, German chemist August W. von Hofmann called it benzene. Twenty years later, benzene's structure was found to be ringlike, with a hydrogen atom attached to each carbon atom. The only way for each carbon to acquire the four additional valence electrons it needs, however, is for there to be three double bonds in the ring.

Later, the ring typical of benzene was found to exist in a group of compounds that smelled good—they were aromatic. The branch of chemistry dealing with benzene-based compounds came to be called aromatic chemistry, even though not all benzene compounds smell good.

47

Organic Chemistry

Benzene occurs naturally in gasoline, but it has been found to be carcinogenic (cancer-causing) to people. Since 1990, the United States government has required that the benzene content of gasoline be reduced. The harmful benzene is removed by reacting it with hydrogen, causing it to form a different hydrocarbon called cyclohexane. Benzene also is used in manufacturing aspirin.

The fractional distillation of petroleum

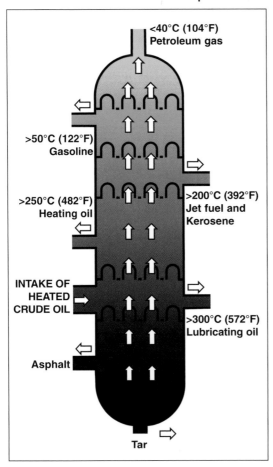

Fractional Distillation

How do we get gasoline out of petroleum? Early in the history of creating chemicals from petroleum, it was discovered that each increasingly complex member of a hydrocarbon family in petroleum boils at a slightly higher temperature than the simpler members. This allows petroleum-refining engineers to separate out each of the different hydrocarbons at different temperatures in a process called fractional distillation.

In the fractionating process, the crude oil as it comes from the ground is pumped into a huge fractionating tower. As heat is applied, different fractions, or kinds of hydrocarbons, begin to boil at different temperatures. As a compound evaporates at a specific temperature, it is piped into a condenser where it is cooled to a liquid again.

One by one, the crude oil's components are removed at different levels within the tower. Gasoline is a mixture of hydrocarbons, with the number of carbon atoms ranging from 5 to 12. Gasoline boils at between 27° and 200°C (80° and 392°F). Jet fuel and lubricating oil fraction out at higher temperatures. Even the remaining "sludge" that almost doesn't boil is used as asphalt for tarring roads.

There isn't really as much gasoline in petroleum as our car-driving society demands. So chemists have learned how to change the chemistry of petroleum. In a process called catalytic cracking, chemists can break large petroleum molecules into the smaller gasoline molecules.

Polymers and Plastics

The ability to rearrange the atoms of hydrocarbon molecules has led to the widespread development of different chemical substances to suit many needs. Often, the rearranging consists of joining many small molecules, called monomers, into long chains, called polymers.

In the polymerization of hydrocarbons, changes are made in the double carbon-carbon bond—two carbon atoms sharing two electrons. In a double bond, the two bonds linking the carbons are not equal—one is strong and one is weak. In linking monomers to make polymers, the weak link is broken. This frees one electron on the carbon atom to be shared with another carbon atom, gradually making a chain of carbon atoms.

The second molecule can then be linked to another one and then another, as shown here:

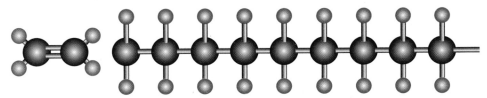

Polyethylene plastic milk jugs that have just been molded. Such plastic can be recycled—melted and reused.

The carbons on each end have a free electron for attaching to more monomers. This is polyethylene, the polymer (consisting of thousands of ethylene—C_2H_4—units) used in making pipes, toys, and lightweight plastic bags.

Today, hydrocarbons play an important role in our lives. However, if they continue to be derived from petroleum, eventually we will run out of this vital natural resource. Perhaps over the next few million years the hydrocarbons developed from organisms living now will create new fossil fuels.

Other Families of Organic Chemicals

In addition to hydrocarbons, other families of organic compounds exist. Methane (CH_4), which is the simplest hydrocarbon, is the starting point of the huge group of organic chemicals. If one hydrogen atom in methane is replaced with the —OH group, or partial molecule, the result is another organic family called alcohols.

If the fourth hydrogen in methane is replaced by a —NH$_2$ group, the new molecule is called an amine (as in amino acids). Another family is the carboxylic acids, which have one hydrogen replaced by the carboxyl group, —COOH.

A CARBON CATALOG

Messy Fiber

Cellulose from trees has wide industrial uses. At least half of it is used in making pulp for manufacturing paper. Pulp is the "mess" that results after wood chips under pressure have been treated by steam and various chemicals. The chips are processed into a fluffy material that is mixed with water. The pulp is then run through a series of heavy rollers that gradually dry it and straighten out the fibers. To make white, high-quality paper, the glue-like material called lignin in the pulp is removed, leaving mostly fiber. Cheaper paper, especially newsprint, has the lignin in it.

The cellulose in processed and bleached wood pulp has many uses.

Musical Graphite

Musical instruments such as acoustic guitars have traditionally been made from hardwoods taken from mahogany, ebony, and rosewood trees. These are the very trees that have been put on the endangered species list. Instrument makers searching for new materials have found considerable success in making acoustic guitars from graphite and from a plastic impregnated with the material used for bullet-proof vests. Musicians find that these materials have an important advantage over wood—the sound produced does not change with heat and humidity, nor will the graphite material crack from getting too dry the way wood might.

Fiber for Strength

In recent decades, fibers from carbon have been used quite differently than they are used in making paper. Thin fibers or "whiskers" of carbon made from heated organic materials can be set in a plastic. This carbon-fiber reinforced plastic, referred to as CFRP, is both lightweight and very strong. This way of using carbon fibers was originally developed during the 1960s for use in building high-performance aircraft. Now such fiber composites are used in making sporting equipment such as poles for vaulting and golf clubs.

Carbon fibers being arranged into a mold to reinforce plastic parts used in extra-strong bicycle components

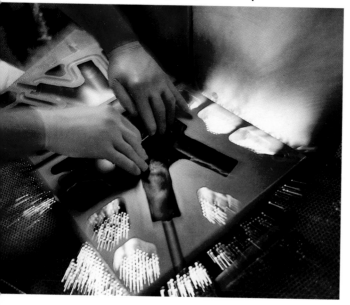

Activating Carbon

Activated carbon is usually charcoal that has been heated to a high temperature (about 900°C; 1,650°F). Impurities that adhere to the charcoal's surfaces are driven off, leaving many large holes, or pores, in the charcoal. The result is a huge surface area that attracts other substances. Activated carbon, often called activated charcoal, is useful for removing impurities and odors from various chemicals.

Activated carbon works by *ad*sorbing the material to be collected or removed from waste. *Ad*sorption means that the molecules of the waste material cling to the surface of the carbon instead of being taken inside it (which would be *ab*sorption). Activated carbon is also used to adsorb the natural brown color found in many liquid foods, such as vegetable oils and liquid sugar.

Cages of Methane

Huge supplies of methane gas (CH_4) are deposited in large pockets below the floor of the ocean. Some of this methane may be trapped within "cages" of frozen ice crystals that are occasionally produced by the cold temperatures and great pressure of the water above. Called methane hydrates, these odd crystals have the ability to contain huge amounts of free gas. It has been estimated that there is more methane trapped in such hydrates than there is in all other petroleum and natural gas sources. Geologist Jerry Dickens suggests that perhaps it was the release of a large number of these hydrates all at once—thus warming the ocean—that created major climate changes in the past.

Lighting up with Carbon

Carbon played an important role in early electric lighting. Thomas Alva Edison, the inventor of the lightbulb, used a cotton thread coated with carbon black as the filament in some of his early bulbs.

The first system of electrical streetlights was built in Cleveland, Ohio, in 1879. But these lights were not like Edison's lights. The system's developer, Charles F. Brush, used carbon-arc

When this type of carbon-arc lamp was installed in Cleveland, Ohio, in 1879, such lights were the world's first electric streetlights.

lights in which light occurs when an electric current jumps the gap in a lamp between a positive electrode and a negative electrode in order to complete an electric circuit. In Brush's lights, the electrodes were made of carbon. Carbon-arc lamps are still used in searchlights because they make such bright light.

Carbon is also used indirectly to provide light. Carbon black is compressed into rods for use in carbon-zinc flashlight batteries. In such batteries, electrons flow from the zinc (Zn, element #30) battery case (which is covered by a manufacturer's label), through the lightbulb, causing it to light up, and then into a carbon rod in the battery. Surrounding the rod is a thick paste of chemicals that take on electrons from the carbon rod and give up electrons to the zinc casing.

Metallic Carbon

A carbide is a metal that has been combined with carbon at high temperatures. The carbon atoms give the metal special qualities of hardness, wear resistance, and even the cutting power of diamonds. Carbides are used in making drill bits and grinders for factory machining processes.

One such compound, silicon carbide (SiC), is better known as Carborundum®. Silicon carbide is a very hard substance. Sand-like bits of it embedded in a stone wheel make a long-wearing grinding stone. Chunks of Carborundum are part of the metal used to make the very hard drill bits needed for drilling for petroleum.

Other carbides are combined with various metals, such as titanium (Ti, element #22), tungsten (W, #74), and chromium (Cr, #24)). The result is "cemented carbide" that can be molded into various shapes. This extra-tough metal produces a machine part that will hold up to abrasion long after even the toughest steel alloy would have worn away.

Tungsten carbide, which has the chemical symbol WC, is the primary metal used in drill bits for drilling oil wells through rock. Without tungsten carbide's superior hardness, the worn, dull bits would have to be changed so often that drilling would be too costly and time-consuming to be profitable.

Hard as Diamond

Because diamonds are so hard, they have long been sought for industrial use. But even industrial-quality diamonds are scarce, which is why they have been synthesized for thirty years. But now a new form of diamond is being developed. This new form of diamond is created by a process called chemical vapor deposition (CVD).

In the CVD process, a carbon-containing gas (usually methane, CH_4) is heated over a solid surface or activated by microwaves. The carbon atoms settle onto the solid surface in a thin film of diamondlike crystals. Scientists hope eventually to use such films as powerful heat collectors, as coatings on cutting or abrasion tools, or in various electronic devices.

Nonstick Surfaces

Chains of carbon atoms to which fluorine atoms are attached make a special type of compound that does not react easily with other chemicals around it. This compound has long been used as a coating on pots and pans to prevent foods from sticking to the cooking surface. Although the compound will not burn, it begins to fall apart at about 600°C (1,112°F).

Saving Artwork

In 1996, scientists at the National Aeronautics and Space Administration (NASA) responded to a request by the Cleveland Museum of Art to find a way to remove carbon soot from paintings after a fire. In processes used previously, the chemicals used to remove soot usually damaged the paint underneath the soot.

The space scientists had observed that the coatings on satellite surfaces were often removed by the single-atom oxygen gas in the upper atmosphere as the satellites orbited through this region. The scientists put a blackened painting in a vacuum chamber containing a gas of single-atom oxygen. Within hours, the oxygen had combined with the carbon of the soot, forming carbon monoxide and carbon dioxide. The soot was gone, and the paint underneath, while dried slightly, was unharmed.

Following the Isotope

In the top layers of our atmosphere, the nuclei of nitrogen atoms are struck by cosmic rays, which consist of nuclei and subatomic particles of other atoms. Occasionally, the nuclei of atmospheric nitrogen atoms absorb extra neutrons.

A common nitrogen atom has seven neutrons and seven protons in its nucleus; it is nitrogen-14. But when an extra neutron is taken into the nitrogen atom's nucleus, a proton may be emitted. If this positive particle is gone, an electron, with its balancing negative charge, is also emitted. Left behind is an atom with only six protons and six electrons, but eight neutrons. The nitrogen-14 atoms have became carbon-14 atoms. Carbon-14 is one of the radioactive isotopes of carbon.

In the late 1940s, American scientist Willard F. Libby determined that the nitrogen-to-carbon process occurring in the atmosphere produces a constant amount of carbon-14 in our environment. Some of that carbon-14 is always present as part of

Soon after carbon dating was discovered in the 1940s, archeologists began to use the process to verify the actual time from which their samples came.

the carbon dioxide that plants and animals take in. When these living things die, the C-14 is detectable by scientific instruments that indicate radioactivity. The concentration of radioactive carbon-14 in an object is reduced by half in 5,730 years—a period called its half-life. The age of an item made thousands of years ago by ancient people, such as a wooden stool or a shell button, can be determined from a measurement of the amount of carbon-14 radioactivity left. The amount of carbon-14 remained constant while the oyster or tree was alive, but it began to decrease when the tree was cut or the oyster killed.

The process of carbon-14 disappearing is the basis of a type of dating used in archeology. Archeologists can use the change in the amount of carbon-14 in objects as a way of determining how old the objects are.

Carbon in Brief

Name: Carbon

Symbol: C

Discoverers: Known since ancient times

Atomic number: 6

Atomic weight: 12.011

Electrons in shells: 2, 4

Group: 14 (sometimes called 4A); other elements in Group 14 include silicon, germanium, tin, and lead

Usual characteristics: Depend on the form, or allotrope. Diamond is a very hard, generally colorless crystal. Graphite is also crystalline, but the atoms in this soft, slippery black material form layers that slide over one another. Amorphous carbon is black and powdery. Buckminster-fullerene does not occur in identifiable masses

Density (mass per unit volume):
diamond: 3.5 grams per cubic centimeter
graphite: 2.26 grams per cubic centimeter.
amorphous carbon: 1.8–2.1 grams per cubic centimeter

Sublimation point (goes directly from solid to gas): Both graphite and amorphous carbon sublime to a gas at about 3,550°C (6,420°F). Diamond changes into graphite before that temperature is reached.

Boiling point: 4,827°C (8,721°F)

Abundance:
Universe: 5th most abundant at 0.021%
Earth: Carbon and several other elements together make up only 1.4%
Earth's crust: 11th in order of abundance at about 0.2%
Earth's atmosphere: Exists as carbon dioxide, carbon monoxide, and methane, totaling less than 340 ppm
Human body: 3rd in order of abundance at 9.5%

Stable isotopes (carbon atoms with different numbers of neutrons): C-12 (98.9%) and C-13 (1.10%)

Radioactive isotopes: C-10, C-11, C-14, C-15, and C-16.

Glossary

acid: definitions vary, but basically an acid is a corrosive substance that gives up a positive hydrogen ion, H+, equal to a proton when dissolved in water; acids measure less than 7 on the pH scale because of their large number of hydrogen ions

allotrope: an alternative structure of an element when it exists in two or more forms. The allotropes of carbon include diamond, graphite, amorphous (meaning "formless) carbon, and (according to some scientists) fullerenes.

anion: an ion with a negative charge

atom: the smallest amount of an element that exhibits the properties of the element, consisting of protons, electrons, and (usually) neutrons

base: a substance that accepts a hydrogen ion (H+) when dissolved in water; indicates higher than 7 on the pH scale because of its small number of hydrogen ions

boiling point: the temperature at which a liquid at normal pressure evaporates into a gas, or a solid changes directly (sublimes) into a gas; also, the temperature at which a gas or vapor condenses into a liquid or solid

bond: the result of the attractive force linking atoms together in a molecule or crystal

catalyst: a substance that causes or speeds a chemical reaction without itself being used up or consumed in the reaction

cation: an ion with a positive charge

chemical reaction: a transformation or change in a substance involving the electrons of the chemical elements making up the substance

combustion: burning, or rapid combination of a substance with oxygen, usually producing heat and light

compound: a substance formed by two or more elements bound together by chemical means

covalent bond: a link between two atoms made by the atoms sharing electrons

crystal: a solid substance in which the atoms are arranged in three-dimensional patterns that create smooth outer surfaces, or faces

decompose: to break down a substance into its components

density: the amount of material in a given volume, or space; mass per unit volume; often stated as grams per cubic centimeter (g/cm³)

diatomic: made up of two atoms

distillation: the process by which a liquid is heated until it evaporates and its component gases are collected and condensed back into a liquid in another container; often used to separate mixtures into their different components

DNA: deoxyribonucleic acid, a chemical in the nucleus of each living cell, which carries genetic information

double bond: the sharing of two pairs of electrons between two atoms in a molecule

electrode: a device such as a metal plate that conducts electrons into or out of a solution or battery

electron: a subatomic particle with a negative charge that continually moves within a region surrounding the nucleus of an atom

electron shell: the region within which electrons with specific energy levels move

element: a substance that cannot be split chemically into simpler substances that maintain the same characteristics. Each of the 103 naturally occurring chemical elements is made up of atoms of the same kind.

evaporate: to change from a liquid to a vapor or gas

fossil fuel: petroleum, natural gas, or coal, all of which are formed from the remains of plants and animals

gas: a state of matter in which the atoms or molecules move freely, matching the shape and volume of the container holding it

group: a vertical column in the Periodic Table, with each element having similar physical and chemical characteristics

half-life: the period of time required for half of a radioactive element to decay

hormone: any of various secretions of the endocrine glands that control different functions of the body, especially at the cellular level

hydrocarbon: a compound made of only carbon and hydrogen

inorganic: not containing carbon

ion: an atom or molecule that has acquired an electric charge by gaining or losing one or more electrons

isomers: two or more molecules that have the same chemical composition but different arrangements of atoms

isotope: an atom with a different number of neutrons in its nucleus from other atoms of the same element

mass number: the total of protons and neutrons in the nucleus of an atom

melting point: the temperature at which a solid becomes a liquid, or a liquid changes to a solid

metal: an element that conducts electricity, usually shines, or reflects light, is dense, and can be shaped; about three-quarters of the naturally occurring elements are metals

metalloid: a chemical element that has some characteristics of a metal and some of a nonmetal; includes some elements in groups 13 through 17 in the Periodic Table

molecule: the smallest amount of a substance that has the characteristics of the substance and consists of two or more atoms

monomer: a molecule that can be linked to many other identical molecules to make a polymer

neutral: 1) having neither acidic or basic properties; 2) having no electrical charge

neutron: a subatomic particle in the nucleus of all atoms except hydrogen; has no electric charge

nonmetal: a chemical element that does not conduct electricity, is not dense, and is too brittle to be worked; nonmetals easily form ions, and they include some elements in groups 14 through 17 and all of group 18 in the Periodic Table.

nucleus: 1) the central part of an atom, which has a positive electrical charge from its one or more protons; the nuclei of all atoms except hydrogen also include uncharged neutrons; 2) the central portion of most living cells that controls the activities of the cells and contains genetic material

organic: containing carbon

oxidation: the loss of electrons during a chemical reaction, which occurs in conjunction with reduction; need not involve the element oxygen

pH: a measure of the acidity of a substance, on a scale of 0 to 14, with 7 being neutral. pH stands for "potential of hydrogen"

photosynthesis: in green plants, the process by which carbon dioxide and water, in the presence of light, are turned into sugars

plastic: any material that can be shaped, especially synthetic substances produced from petroleum

polymer: a large molecule formed by the repeated linking of small molecules, or monomers.

pressure: the force exerted by an object divided by the area over which the force is exerted. The air at sea level exerts a pressure of 14.7 pounds per square inch (1013 millibars).

protein: a complex biological chemical made by the linking of many amino acids

proton: a subatomic particle within the nucleus of all atoms; has a positive electric charge

radical: an atom or molecule that contains an unpaired electron

radioactive: spontaneously emitting high-energy particles

reduction: the gain of electrons, which occurs in conjunction with oxidation

respiration: the process of taking in oxygen and giving off carbon dioxide

salt: any compound that, with water, results from the neutralization of an acid by a base. In common usage, sodium chloride (table salt).

shell: a region surrounding the nucleus of an atom in which one or more electrons can occur. The inner shell can hold a maximum of two electrons; others may hold eight or more. If an atom's outer, or valence, shell does not hold its maximum number of electrons, the atom is subject to chemical reactions.

solid: a state of matter in which the shape of the collection of atoms or molecules does not depend on the container

sublime: to change directly from a solid to a gas without becoming a liquid first

synthetic: created artificially instead of occurring naturally

triple bond: the sharing of three pairs of electrons between two atoms in a molecule

ultraviolet: electromagnetic radiation which has a wavelength shorter than visible light

valence electron: an electron located in the outer shell of an atom, available to participate in chemical reactions

For Further Information

BOOKS

Atkins, P. W. *The Periodic Kingdom: A Journey into the Land of the Chemical Elements.* NY: Basic Books, 1995

Heiserman, David L. *Exploring Chemical Elements and Their Compounds.* Blue Ridge Summit, PA: Tab Books, 1992

Hoffman, Roald, and Vivian Torrence. *Chemistry Imagined: Reflections on Science.* Washington, DC: Smithsonian Institution Press, 1993

Newton, David E. *Chemical Elements.* Venture Books. Danbury, CT: Franklin Watts, 1994

Yount, Lisa. *Antoine Lavoisier: Founder of Modern Chemistry.* "Great Minds of Science" series. Springfield, NJ: Enslow Publishers, 1997

CD-ROM

Discover the Elements: The Interactive Periodic Table of the Chemical Elements. Paradigm Interactive, Greensboro, NC, 1995

INTERNET SITES

Note that useful sites on the Internet can change and even disappear. If the following site addresses do not work, use a search engine that you find useful, such as Yahoo:

> http://www.yahoo.com

or AltaVista:

> http://altavista.digital.com

A very thorough listing of the major characteristics, uses, and compounds of all the chemical elements can be found at a site called WebElements:

> http://www.shef.ac.uk/~chem/web-elements/

A Canadian site on the Nature of the Environment includes a large section on the elements in the various Earth systems:

> http://www.cent.org/geo12/geo12/htm

Colored photos of various molecules, cells, and biological systems can be viewed at:

> http://www.clarityconnect.com/webpages/-cramer/PictureIt/welcome.htm

Many subjects are covered on WWW Virtual Library. It also includes a useful collection of links to other sites:

> http://www.earthsystems.org/Environment/shtml

INDEX

Earth 26, 32, 43, 58
Earth's atmosphere 8, 22, 27, 33, 58
ebony 51
Edison, Thomas Alva 17, 53
egg 41
ethyne 47
electric circuit 54
electric current 53
electric lighting 53
electrical charge 6, 19
electricity 21
electrodes 54
electron rings 6
electron shell 6, 7, 11, 12, 58,
electrons 6, 7, 12, 13, 16, 19, 22, 38, 44, 46, 47, 49, 50, 54, 58
element groups 7
element name 6, 58
element number 6, 11
elements 6, 7, 8, 9, 10, 11, 12, 13, 22, 24, 32
endangered species 52
energy 28, 31, 33, 34, 37, 41
energy levels 7
enzymes 37
ethane 45
ethylene gas 47
exercise 39
Expo '67 World's Fair 20

fats 35, 37, 50
fatty acids 37, 38
fermentation 10, 25, 35
fibers 36, 52
filament 53
fire 13, 31
fire extinguishers 27
fires 9
fish oils 37
"fixed air" 10
flooding 29

fluids 35, 38
fluorine 12, 13, 55
forests 29, 31
fossil fuels 14, 29
fractional distillation 48
free radicals 44, 45
fructose 34, 35, 36
fruits 34, 35, 47
fuels 34, 47
Fuller, Buckminster 19, 20
fullerenes 15, 21, 22
fullerite 21

galactose 34
gases 7, 10, 13, 19, 23, 24, 25, 46, 47, 53, 55
gasoline 29, 48, 49
General Electric Co. 17
genes 42
genetic engineers 42
genetic material 41
geodesic dome 20
Georgia Institute of Technology 21
germanium 58
global warming 28
glucose 33, 34, 36, 37
glycine 39, 40
glycogen 35
gold 11
golf clubs 52
grains 35
graphein 16
graphite 9, 10, 15, 16, 17, 18, 21, 52, 58
graphite rods 18
grass 36
greenhouse gases 28
grinders 54
group 45
guanine 41

hair 39
half-life 57
halogens 13

hardwoods 51
HDL, see lipoprotein
heart 38, 39
heat 10, 15, 16, 52
heat waves 28
Heath, James R. 8
heating 44
helium 22
Helmont, Jan Baptista van 9, 10, 23
hemoglobin 30, 31
human body 34, 37, 38, 39, 40, 42, 58
humans 29, 32, 34, 44
hydrocarbons 14, 43, 44, 45, 46, 47, 48, 49, 50
hydrogen 6, 11, 12, 14, 33, 43, 44, 46, 47, 48, 50
hydrogen atom 38, 40, 44, 47, 50

Ingenhousz, Jan 24
ink 19
internal combustion engines 29
ions 19
iron 10, 30
isomers 45
isotopes 6, 12, 56, 58

jet fuel 49

K shell 7
Kroto, Harry 19

L shell 7
lactose 34
Lake Nios 26
lampblack 18
lanthanide elements 7
laser 19
Lavoisier, Antoine 10, 24
LDL, see lipoprotein

lead 12, 16, 58
leaves 33
Libby, Willard A. 56
light 28, 54
light bulb 53, 54
lignin 51
limestone 14
linoleic acid 37
linolenic acid 37
lipids 37, 38
lipoprotein 39
liver 34, 35, 39
lubricant 16
lungs 30, 31

mahogany 51
mammals 43
Mars 14
mass 6
mass number 12
matrix 15
matter 19
meat 41
Mendeleev, Dmitri 11
metabolism 35
metalloids 7
metals 30, 46, 47, 54, 55
meteorite 14, 22
methane 19, 28, 43, 44, 45, 47, 50, 53, 55, 58
methane hydrates 53
microcircuits 21
microorganisms 36
microwaves 55
milk 34
mineral 15, 16
mineral oxide 30
Mohs Hardness Scale 16
molecules 8, 19, 21, 22, 28, 30, 32, 33, 39, 40, 41, 53
monomers 49, 50
monosaccharide 34
Montreal, Quebec 20
Morveau, Guyton de 10